A "SIMPLY SCIENTIFIC" VALENTINES

Romance for nerds

By Gary McCallister

I0490258

ISBN:

A "Simply Scientific" Valentines

DEDICATION

To my wife Gaydra, although my love
is no laughing matter

CONTENTS

ACKNOWLEDGMENTS

Most of the people responsible for this series have declined to be identified.

1. REALITY

Humans tend to be confused about reality. Scientists, on the other hand, are all about reality. The question becomes, "Are scientists human?"

Well, I think they are. Because I have discovered they are just as confused about romance as everyone else. This might seem like an extravagant claim. What!? Which claim? That scientists are human, or that scientists are confused about romance?

Of course people are confused about reality or else they deny it. C. S. Lewis pointed out that humans sometimes say things like, "That wasn't real. It was just a feeling you had because of the music, the lights, the company, and the wine." In this case, the word "real" refers to an actual physical experience and feelings are just a bunch of endorphins or related hormones.

On the other hand, someone might also say, "You don't understand what it was really like to be hanging out of a second-story window and being shot at." In this case, reality is the feelings one has while being shot at while hanging from a second-story window.

Why were they hanging from a second story window, I wonder? Did it have anything to do with the music, lights, and wine? By the way, the quote marks above simply indicate the things people might say. They aren't direct quotes from C. S. Lewis who, I am sure, made the point more eloquently.

With this kind of loose semantics though, how is a person to know how they are really feeling? When someone asks, "How are you?" I'm thrown into a vortex of confusion. If I don't reply that I am good does it mean I need to repent? Is feeling good the same as feeling well? Why are feelings not real if they are caused by real things like endorphins?

Personally, as pleasant as some feelings may be, I'm not sure I trust them all that much. Think about it. Sometimes I feel happy. Other times I feel sad. Sometimes I feel like working hard. Most of the time I'd rather write a column or a book.

Sometimes I feel incredibly patient with idiots. Other times I don't feel very patient with everyone because everyone else is an idiot. Feelings come and feelings go. Life, based on feelings, doesn't seem to be a consistent way of living.

I guess this is why science isn't very useful on Valentine's Day. One can identify, and even quantify, endorphins. However, their actual affects seem unpredictable and mysterious. One needn't look far to see the baffling results. For example, my wife is intelligent and attractive, but she seems to love me. At least she's still around.

Valentine's Day is all about feelings, which is why there is some question about whether scientists are human. This may not seem urgent to most of you, but I can assure you that on Valentine's Day the smart scientists should try to act like feelings are real even if he or she is confused about the matter. I know this seems to be much clearer to non-scientists.

I think a better basis for living life is simply to make love a decision. Then if the feelings change, the obligations one undertakes in the name of love don't. "I've decided to be a good husband to you forever" probably doesn't sound too romantic. But it's a lot more reliable than saying "I will love you forever." Because I must continue to be a good husband even if I don't feel like it sometimes.

What is interesting is that one of the best ways

of changing our feelings is simply to decide to change our feelings. If we wait until we feel like changing our feelings, we may never feel like it. Or we might feel like changing the feeling of changing our feelings. Then where would we be? I have no idea!

The questions remain. Are feelings real, and are scientist's human? No, wait! Was it, are scientists real and are feelings human? I don't think I feel like talking about this anymore.

2. THE CHEMISTRY OF ROMANCE

I think Chemistry has something to teach us about Valentines. I know this is a little late for whatever trouble you got yourself into last night. That's the problem. Love always seems to make things more complicated. Sometimes it's in a good way, but it's still more complicated.

This is where I think Chemistry can help. I should clarify that this is my kind of chemistry, and real chemists might not agree. In fact, if you are a real chemist, why don't you skip to the sports page?

Chemistry is the study of fundamental particles, called elements and how they interact with each other to form more fundamental particles called molecules. Molecules interact with each other to form even more complicated molecules called by different names depending on how they interact. These more complicated molecules interact to cause explosions and other, less interesting, phenomenon. See? Right there you can see all kinds of analogies to the world of romance.

The basis of chemistry is how two different elements interact with each other. I am not sure if men and women are different elements, but I am pretty sure they are different. Is that politically incorrect? Anyway, elements can basically interact in two ways. Chemists will tell you that it's a little more complicated than that, but that's what chemists always say. Come to think of it, that's what marriage counselors always say too.

One way elements interact is by having an electrical charge that attracts another element with the opposite electrical charge. As everyone know, opposites attract. When these two elements get close to each other they are attracted in the same way magnets are attracted to each other. They form a magnetic bond called an ionic bond. This is what happens when sodium and chlorine get close together and form salt.

The problem with ionic bonds is that there are lots of electrically charged atoms and molecules out there. If one with a stronger positive charge comes by, it can pull the negatively charged atom away and vice versa. This leaves the first positively charged atom broken hearted. Well, at least lonely until it can find a new negatively charged atom. Anyway, salt is a good analogy, even if it is poured into the wound.

Then there are the second way elements can bind together. Some atoms have spare electrons floating around. Other atoms often have room for an extra electron or two. Sometimes such atoms decide to share electrons. One atom keeps the electron for a while, then the other atom takes it back. Neither are exactly content all the time, but both are happier than they would have been without the other.

These are called covalent bonds, and they are much stronger than ionic bonds. Think of it like two brothers who buy a motorcycle together. They constantly argue about who gets it when, but they both get to be cool more often than if they hadn't pooled their funds. That's called a brotherly bond. Those two brothers stay close because they share a motorcycle. (That would be at least until one gets drafted and the other sells the motorcycle keeping all the money for himself. But that's all forgotten, and it is seldom even mentioned now.)

The point is, sharing things together makes a much stronger bond that just being attracted. My wife and I share four different wedding ceremonies (I'd explain, but it really isn't very scientific). That alone would complicate any separation since I suppose we would need four different divorces. Who could afford that? Why else do you think she sticks around?

Of course, in addition we share a house, a car (she says the old truck is all mine) and a long history that includes four children, their spouses, and 18 grandkids. So in general, the more things you share, and the more equal the sharing, the stronger the bond, and the less likely it is to be disrupted. Sounds like romance to me.

3. WHAT ARE PEOPLE FOR?

Since my retirement from the University, I no longer have large research grants and extensive laboratory space for my research. Well, I've never had any large research grants, and most of my research had to be done in the basement of the old Wubben Hall because all lab space was needed for classwork. Oh, yeah, there is a basement in the old Wubben Hall. You don't want to go there.

Anyway, I now have time to speculate on some less-important scientific questions that have defied analysis. Like, "What are people for?" I am always being asked, "What are mosquitoes for?" or, "What are centipedes for?" I think what people mean is, "What good are mosquitoes and centipedes to people?"

I have read that some people think there are too many people. How can we decide there are too many people unless we know what people are for? If the job people are for is getting done and there are people with nothing to do, then I suppose we should get rid of a few of them. I have a couple of suggestions.

I'm reminded of the time, several years ago, when the government was telling us that there were too many farmers. We had to get rid of some of them, so we could be more efficient. Interestingly, I never heard a farmer express that concern. Too, I never heard a professor of agriculture opine that there were too many professors of agriculture. However, if we had fewer farmers, wouldn't we need fewer professors of agriculture, and a smaller Department of Agriculture?

Where did all those farmers go anyway? They went to cities of course. Apparently, our cities need a lot more people than our farms to do whatever "people are for". I sort of know what people are for on a farm. I am not as sure why we need people in the cities, but there must be some reason because all the farmers went there.

In 2014 it is starting to look like we have too many people in the cities. With all the efficient farms, automated factories, labor-saving devices, robots, shorter workdays, longer vacations and early retirement, it sort of appears that people are just for goofing off. My wife thinks that's what I've been doing for years. . .

Maybe there are too many people in the cities and on the farms both. If we don't know what they're for, it's hard to tell. Of course, if we think there are too many, I suspect there are further inhumanities on the horizon.

Then again, I may be going about this all wrong. Maybe I should be concentrating on the question, "What good are people to other people?" If we judge other living creatures by this standard it seems only fair to judge people in the same way. Several testable hypotheses come to mind.

If it weren't for the people who invented computers, I would still be writing my columns in long hand. Therefore, my editors would be unable to read my columns, so you would never have to suffer through this stuff! Okay, I guess that could be taken as a pro or a con.

But without people, we wouldn't have electricity, cars, airplanes, indoor plumbing, television, cell phones, or Facebook. Nor would we have Shakespeare's plays, Rembrandt's paintings, or Beethoven's symphonies. Ideas and creations come from people. In fact, as far as I know, ideas never come from anything but people.

There is at least one other thing that humans "might be for". Almost all people are fellow humans, friends, neighbors, parents, children, citizens, children of God, and kind people who read my newspaper columns and books. So my scientific hypothesis is, people are for taking care of each other through caring, creating, and having better ideas. That means I just must figure out what all those people who don't read my books are for.

4. SCIENCE FICTION AND FEELINGS

My wife doesn't like science fiction. She thinks it's unbelievable. Yet, she watches the news and listens to State of the Union speeches. I find science fiction interesting. In fact, when I die, I think I'll leave my body to science fiction. They'll probably do more with it that I ever have. I think it's fascinating that there is even a genre like science fiction. I mean the words "science fiction" must be an oxymoron, right?

Interestingly the term "science fiction" didn't exist until about 1929 when Hugo Gernsback coined the word for his new magazine. He first used the term "scientifiction" for stories that he defined as, ". . . the Jules Verne, H.G. Wells, and Edgar Allan Poe type of story." The awards given out annually by the World Science Fiction Convention are called "Hugo's" in his honor. I think getting one is supposed to be an honor.

Prior to Gernsback, such stories were called "scientific romances". That term apparently originates about 1845 when it was used to describe Robert Chambers' "Vestiges of the Natural History of Creation". You're probably familiar with that publication. Even H. G. Wells used the term "scientific romance".

Apparently, in earlier years, the term "romance" was more or less synonymous with fiction. I know, crazy, huh? Today, of course, romance deserves its own genre and usually has little to do with science. I thought, briefly, about reinventing the concept of scientific romances, but I think trying to make scientists realistically appear romantic is a bit of a stretch. Readers can't generally set aside that much unbelief.

It's not that scientists aren't romantic. Scientists

have feelings too. Well, that's the theory anyway. It's just that they are generally kind of awkward when it comes to romance. I am not sure if that generalization is true or not as I have not seen any reliable scientific data on romance involving scientists. Actually, I don't think there is any research on the subject. I suspect funding agencies would find the idea of doing research on romance among scientists too romantic to be science, if you know what I mean. This is extremely shortsighted, however, as the number of scientists in the world is seriously declining.

But here's the problem. When a "science kind of person" falls in love, they simply have not developed the tools for dealing with the situation. The layperson thinks, when scientists make a discovery, they shout something like "Eureka!" In truth, when scientists discover a new phenomenon like an attractive person, they usually say, "Wow, that's weird!" This response to someone of the opposite gender can be confusing to them.

Further, things like logarithms, functions, fractals, and chi square are not particularly useful in romance, and the science-minded may have to exhaust these approaches before they recognize they are dealing with a new situation. Be patient.

In fact, science fiction and romance have a lot in common. Science fiction, like romance, is about an idea that doesn't exist yet, but soon could. When it does it will change everything for everybody, and nothing will ever be the same again. That happened to me once. Luckily, it was back when I thought I was a poet, and before I became a scientist, so it was a lot easier for me. Or maybe it was easier for my wife.

That does raise interesting questions about when one becomes a scientist and when one has officially fallen in love. Neither destination has a clearly marked designation. I knew I was in the early stages of love quickly. However, I didn't know I was a scientist until someone addressed me as Dr. McCallister. I thought she was talking to someone behind me with the same name. I've also always believed that Mr. McCallister was my Dad.

As that great science fiction writer Henry David Thoreau, observed, "Things do not change. We change." Some people think romance is a fantasy. But romance definitely falls into the science fiction category because science fiction is about improbable possibilities and romance deals with plausible impossibilities.

5. WEDDINGS

It's nearly June which leads me to reflect on weddings. Actually, I didn't choose to reflect on weddings. It was forced on me by the cascade of articles and advertisements about weddings this time of year. I was also thinking about chromatography, but weddings seemed more seasonally apropos.

Darwin claims in his book The Descent of Man, published in 1871, that "Males of almost all animals have stronger passions than females . . . Females want mates who are vigorous and well-armed." He based a lot of his theory of sexual selection on the peacocks' tail. It isn't clear, however, if female peacocks share our fascination with male peacock tails or if the tail is a particularly vigorous weapon.

This is important because of the so-called war between the sexes, not to mention the war on women. According to Darwin, the two genders want different things: females want mates with good genes while males want a lot of sex. This theory suggests there is a necessary, immutable, and inevitable conflict between the sexes.

Now I just flat out disagree with that. No offense to peacocks. While this gender concept is widely taught in biology classes and propounded in modern culture there is almost no scientific evidence to support the idea. Of course, turning to scientists for romantic advice is always ill-advised. Take my word for it.

Now it's tempting to believe that females always choose the best-looking males simply because I hope my wife did that. Of course, I believe that because that is how I chose my wife. However, there is no evidence that handsome, male peacocks are good fighters, territorial kingpins, faster than greased lightning, or smooth talkers either. There is even less evidence that I am any of those things.

The biggest problem with Darwin's theory of gender roles is that it simply doesn't describe what exists in the animal world. For example, there is little evidence that the females of many animals even care with whom they mate. How about herd animals where the winner between the males takes all the girls. In many animals, mostly invertebrates, huge numbers of eggs are deposited in the environment and any old male does fine.

Did you know there are a lot of animals that don't even require males for mating? Virgin births are possible in some species. In fact, most animals have a very limited number of opportunities to mate either because of low populations or the time constraints that arise from seasonal reproduction.

Additionally, in many species, the behaviors are reversed with males nurturing the offspring. In other species, the two genders look exactly alike. In still other species, the females are the more colorful. There is even little evidence that females are less passionate. Then what about the species that undergo sex changes as a normal course in their life cycle, changing from male to female or vice versa? So no, humans did not invent transgender issues.

The problem with Darwin's theory is that it has created a thousand generalizations, two thousand exceptions, and numerous special-case scenarios that must be explained away. That isn't a theory. It's an idea that every exception disproves.

It seems to me that biologically, if not romantically, males and females begin with the exact same goal. They want to have offspring. I know. That's sort of a crazy theory in these modern days. But from the beginning, males and females have been involved in a joint venture. When we examine actual animal behavior, we see far more cooperation than conflict.

Well, I probably should have written about chromatography. I probably know more about that. Chromatography is a method of separating mixtures. Weddings appear to be more about bringing separate things together. It appears that there is very little actual science involved in planning weddings. But the creation of children is the study of biology by synthesis. Personally I have been a practicing biologist. Many of my colleagues are just theoretical.

6. UTILTIY COSTS

It's February, so let's consider the actual value of the heart in terms that make sense to everyday people.

Let's see, the water service takes water out of the river, that came from the snowpack, that originated in the sky, that came from ocean water evaporation. The service cleans it up a little and pumps it to your house where you use it. Then, it ultimately flows back to the river and the ocean. That's called the water cycle.

Obviously, the water district didn't invent water or manufacture it. All it really does is add some value in the form of safety from disease and convenience. For their service they charge a small fee of about $0.005 cents a gallon per month. However, if you exceed the basic amount of water allotted by water companies per day, about 3000 gallons, you will have to pay about $0.002 cents per gallon more each month.

That seems reasonable. Especially if you consider that a bottle of water in the store can cost a dollar or more. A sixteen-ounce bottle of water, at that rate, is about $9.00 per gallon.

For comparison, consider that your heart pumps about 2000 gallons of blood per day, or 60,000 gallons of blood per month. While distances are less, it's still a recycle of the same stuff over and over again. In addition, your body has the added responsibility of creating the components of the blood stream.

When comparing the two you can't even argue that the water company provides value added in the form of clean up and convenience. Your liver performs the same "decontamination service", and you can't get much more convenient than a heartbeat. Mine hasn't missed a beat in years. I think it's even more reliable than my newspaper carrier and I thought he was pretty good.

If I had to pay my heart at the same rate I pay the water company, it would cost me $11.00 on the first day of the month. By the second day, I'd owe another $5.50 plus $2.35 for the extra one thousand gallons. The third day would cost me another $4.70. The next three days would cost an additional $11.75. That would leave me twenty-four more days each month at the approximate rate of $3.25 per 1000 gallons, or $78.00 dollars. My monthly bill for circulating my blood would be $113.30 cents.

Of course, these are all just assumptions based on fair practice in water districts and standard, healthy bodies. Assuming either "fair practice" or "healthy bodies" may be dangerous assumptions.
The actual amount of blood pumped per minute can vary from about one gallon to ten gallons per minute, depending on whether your wife is angry at you or not. However, this all averages out over time since women's hearts beat about 78 times per minute while men's only average about seventy. Some see this discrepancy as an indication that women are perpetually angry at men, but there is little evidence to support this idea.

If you like logistics, you might find it interesting to know that a blood cell only takes six seconds to go from the heart to the lungs and back. It supposedly takes eight seconds for blood to go to the brain and back. This probably explains why I can breathe faster than I can think. My toes have always been the dumbest part of me, and this makes sense because it takes blood sixteen seconds to make that round trip.

So, you can see, that you are really getting your money's worth out of your heart. I think $113.30 a month is a real bargain considering you are pumping the blood through about 60,000 miles of blood vessels. This seems like a better deal than even the United States Postal service. I think you should buy your heart a box of cherry chocolates!

7. UNITY

I was a soldier from 1966 until 1968. I left the military with mixed emotions: joy and gladness. Don't misunderstand me. I am proud that I served my country, and the experience was very beneficial personally. It's just that I am not really very strong, courageous, or obedient. I think, by the end of two years, both the U.S. Army and I had come to grips with that. We parted ways on pleasant terms.

I mention this only because it illustrates an important point often missed by scientists. Since science is the study of our material world, scientists tend to be pathologically focused on the material part of physical objects. Now, it is the nature of physical objects, so science tells us, that no two objects can occupy the same space at the same time.

Therefore, the study of matter is the study of diversity. When cataloging one element, one force, or one animal from another, we are breaking the world up into smaller, separate segments. Categorization has been useful in many ways, yet harmful in others.

By contrast, many important ideas and concepts are abstractions. Abstractions are not physical objects that can be held in your hands. So abstractions can exist in the same space at the same time. Some people say abstractions aren't real or important. Yet love, democracy, and beauty are very important, but are abstractions.

For example, I experienced both joy and gladness upon discharge from the military. You probably have the same mix of feelings every Friday in the late afternoon. People can be curious and appalled simultaneously. I have even felt free and guilty simultaneously!

Many abstractions accompany each other at the same time and in the same place. Young lovers are often excited and confused simultaneously. I am assured that such emotions as love and irritation can occur simultaneously in wives. I have heard this from a number of sources although some are more trustworthy than others.

Another important limitation to the study of the material world is that the same material object cannot exist in two different places at the same time. However, the same material object can occupy the same space at two different times. In fact, this latter idea is obligatory.

Luckily for reality, abstractions do not have these same limitations. In fact, the situation is almost reversed. Two abstractions can exist at the same time in different spaces. How else could one explain romance as being two different bodies feeling attraction and affection towards each other simultaneously. What about beauty and sadness?

Unlike the material world, it is not necessary for two abstractions to occupy the same space at different times. Abstract ideas can move about from time to time on their own. Abstractions such as "emotions" don't just sit there in the same place all the time. Love can be lost or, at least, temporarily misplaced.

Anyway, I think all the attention being paid to diversity in our world is the result of the scientific revolution where everything has had to be categorized into separate physical spaces. So humans have imbued abstract ideas like race, culture, political parties, and religions with the material characteristic of separateness.

Doing this is detrimental to achieving unity of ideals and goals because unity, an abstraction itself, can apparently only happen in the strange, spiritual world of abstractions. Only in the non-material world of abstractions can ideas such as shared feelings and attitudes unite at the same time and space or occur at the same time in two different places. Perhaps this points to one of the roles that religion can play in our modern world. If we don't see religion as a material that separates, perhaps we could see it as a unifying force.

Abstractions such as love, unity, commitment, and faith, are things that unite man and wife and parent and child, as well as communities and countries. More love!

8. MORE LOVE

Valentine's Day is such an enigma. The emotions of the entire tradition are absolutely based on scientific principles and events, yet scientists, themselves, seem curiously inept at the practice. You would think it would be the other way around.

For example, someone in love might say something like, "How do I love thee? Let me count the ways." Now this counting thing would seem to be right up a scientist's alley. In fact, any good scientist should be able to not only count the ways, but also calculate the mean, standard deviation, and the probability coefficient. At the very least a scientist should be able to do a chi square analysis comparing this love with the love they have for some other person.

However scientists seem generally incapable of counting the ways they love someone. In fact, they frequently either ignore Valentine's Day or completely forget about it. It is difficult to tell the two responses apart, since they appear identical. It also seems that either would be nearly impossible to do considering the massive advertisement campaigns based on the economic interests of the day. Let alone the elevated expectations accompanying the season.

Here is another example. What is so hard about telling a young lady, "O, my luve's like a red, red rose"? Spelling is often not a scientist's strong suit, true. But where would Robert Burns be today without the metaphor of an actual, scientific object like a rose? Well, in all fairness, I guess he'd still be dead. But you know what I mean. Who could be better at making a rose metaphor than a good botanist?

Even the layman knows that roses are

beautiful, smell good, and have nasty thorns. However, a good scientist could also explain that the flowers of most rose varieties, have five petals, and that the number five is the number for a marriage because it is comprised of the female number, two, with the male number three.

Instead, scientists tend to go off on the rose being a woody perennial with hundreds of species and thousands of cultivars. They explain that showy flowers are just a ruse to invite pollination and that roses possess sharp thorns as defense mechanisms against being harmed. All of which emphasizes anything but the romantic aspects of roses. It's like scientists are tone deaf or something.

Lord Byron was not a scientist, but a profligate romantic, who coined the lines, "She walked in beauty, like the night of cloudless climes and starry skies." Now, explain to me why Copernicus, or at least Kepler, didn't come up with those lines. For crying out loud, were there no fair damsels in those days? It seems such a waste.

Astronomers have been studying cloudless skies and starry nights for a long time. I doubt Byron even knew the names of any constellations. He fails to mention even one by name in his entire poem. It seems like there is some kind of flaw in scientists that inhibits their ability to see the essential beauty of their subjects and make application, metaphorically, to emotional matters.

My theory is that this inability is not entirely a genetic defect. Admittedly, some people are seemingly born lacking the metaphor gene. But I think long years of indoctrination through the study of science may also play a role. For example, back when I was an English Literature major playing and singing old-English-folk songs in coffee shops, I was far more romantic than I am today after twelve years of higher education.

Otherwise, it is simply unexplainable why scientists, in the last century or so, haven't come up with something like:

My love exists at the border
Of the space-time continuum,
Forever growing yet infinite,
The original angio-spasm.

So I call on scientists everywhere to rise up to meet the needs and demands of the modern world. Whatever the question, the answer is "more love".

9. BETWEEN TWO MIRRORS ON VALENTINES

I usually poke a little fun at scientists on Valentine's Day. Or maybe I make fun of Valentine's Day by talking about science. Being a scientist, I'm not exactly sure which it is. The two concepts seem to be incongruous somehow. But cold logic and red-hot emotion can make for interesting and amusing columns. Only this year feels different. . .

Last week I wrote about broken mirrors. This week I had the opportunity of standing, with my wife, between two large facing mirrors. A peculiar phenomenon results as one sees his reflection in mirror behind him into the mirror before him, and vice versa. The images recede to seeming infinity in both directions.

I'm sure there are others who have had this experience. But considering how rare mirrors have been in the past and how many people on earth today probably don't have ready access to mirrors, it seems the number of people who have seen such images must be small. Standing between these mirrors, with my wife, made it meaningful in a Valentines sort of way.

It put me in mind of how it must feel to see the earth from space. An astronaut has the perspective of great distance. The mirrors give the impression of time perspective. It's like being able to see yourself far into the future as well as being able to look back into the past.

Standing in front of facing mirrors I imagined the receding images as generations yet unborn. If I concentrate on the images projected forward, each reflection of oneself becomes smaller and smaller, just as our genetic influence also recedes. I wonder how much influence my example and life will have on generations yet unborn.

My wife and I fell in love and had four children. Now we have eighteen grandchildren. Who will they fall in love with? How many great grandchildren will there be? Can we envision our offspring "as numerous as the sands of the sea" as Abraham was promised? What are the genetics involved? What influence will we have had? Does anyone worry about these things anymore?

As I look into the mirrors, I can also see images receding behind. It makes me wonder what influence my great-great grandfather, who died in the Civil War, has had on my life. How am I like him? He fell in love, married and had a son he never saw because he was killed in that war.

I knew a great grandfather born in 1864. He died when I was fifteen. He fell in love with my great-grandmother, Sadie, and they had several children. I remember him "rattling the bones" and singing a love song to my great grandmother on their sixtieth wedding anniversary. That sure had an influence on me.

Valentine columns have always seemed like an easy place to get a good laugh until I stood between the two mirrors. Suddenly the genetics of the past and future seemed pretty connected to falling in love and Valentines. The possible influences of my life and behavior on future generations seemed to matter a little more. The rough and tumble world of western Colorado, back over five generations, suddenly seemed more meaningful too.

Many years ago, on that mountain over there,
A young man sought a wife, his life to share.
She was poor and pretty and he was pretty poor.

She offered him fresh bread when he came to the
door.
He took her for his wife and gave to her his name,
And taught his son to play the poet game.

New branches grow from fallen trees.
The roots are the same ones that once fed me.
So, lift your head and dry your eyes.
It's just another day to be born and die.
You've got the blood, you've got the name,
Now teach your sons to play the poet game.

10. CROSSED WIRES

Imagine you have built a small robot car that is powered by two electric motors, one to each rear wheel. If the right motor revolves more rapidly than the left motor, the car will veer to the left. If the left motor is faster than the right the car will turn right.

Imagine this car has two light sensors on the front of the car, set several inches apart. These light sensors are connected to the motors of the car and control the power to the electric motors such that the more light that hits the sensor the faster the motor turns. The right sensor is connected to the right motor, and the left sensor is connected to the left motor.

Imagine we have placed this car in a darkened gymnasium. It will not move because there is no light. But we have placed a remote-controlled light bulb in the center of the floor. When we turn the light on the car will begin to move. However, because of the distance between the two sensors, the amount of light striking the right sensor will be greater than the amount of light striking the left sensor. This will cause the right motor to revolve faster and the car will veer away from the light until it is exactly facing away from the light so that the amount of light to each sensor is equal. It will also go as far away from the light as possible until the sensors are no longer stimulated.

Imagine you are observing this with a friend from the rafters of the gym. Your friend might say something like, "Wow, that thing really doesn't like the light. It runs and hides. How did you make it do that?" Of course, it doesn't "like" or "dislike" anything. It's a robot. It just appears to be a little like a cockroach.

Stay with me here. This is very applicable to

you.

Imagine you make one small change in your robot; you connect the right sensor to the left motor and the left sensor to the right motor. Then you turn the light off, reposition your robot in the gym, and you resume your perch in the rafters.

When you turn on the light the robot moves, but this time it turns toward the light because the sensor on one side drives the motor on the opposite side. Your friend says, "Oh look, it likes the light and is moving towards it." But wait, something is drastically wrong. As the robot gets closer and closer to the light, each sensor gets more light, and this makes each motor go faster. The robot hurtles directly at the light with increasing speed. You friend screams, "Look out! It's attacking!" as the robot hurtles into the light demolishing light and robot in one grand violent act. "Wow!" Your friend observes after a stunned silence. "That robot really hates the light."

Imagine you painstakingly reassemble your robot. This time you add one more tiny change: a governor on the light sensors so that it increases speed until a certain light intensity is reached. Above that intensity the robot turns off the motor it is wired to.

Meanwhile, back in the gym, this time the robot turns towards the light and rushes towards it as before, but as it gets close it slows, and stops, and sits staring adoringly at the light bulb, never moving a motor. Your friend observes, "Oh look, it's in love with the light."

What has this got to do with you and me? Maybe nothing. But cockroaches, and most insects, have brains that connect to the same side of the body (muscles as motors) as their sensors, whether those are eyes or antennae. You and I have crossed nervous systems. The left brain controls the right side of the body and the other way around. Does that partly explain human aggression? And is the difference between love and violence a simple breakdown of the speed governor, the braking system?

I don't know. But if you think these ideas are intriguing you would enjoy the book "Vehicles: experiments in synthetic psychology" by Valentino Braitenberg, available from MIT Press and on Amazon. But please finish this one first.

11. GEEK ROMANCE

Have you heard about the two, red-blood cells that loved in vein? Well, OK, it's a dumb joke. But don't let me catch you repeating it then! The truth is I don't understand anything about the science of love. I think there are some subjects which just don't lend themselves to scientific explanations.

We scientists need to recognize our limitations.

For example, I have never understood what it is that women find attractive in men. Honestly, what could any woman possibly see in a handsome, smiling, muscled, tanned, kid wearing tight jeans and driving a brand new 4X4 jacked-up pickup? He's probably not older than twenty-five, immature and doesn't have a brain in his head.

Girls, any guy with a great tan doesn't have a job!

Then what's with the sinister dudes? Why are women so surprised when they turn out to be, well, sinister? And don't even get me started on bass players in rock and roll bands.

If you girls really want true romance, you have overlooked a quiet group of substance, stability and culture. Well, OK, it's a unique form of culture, but it is one. I'm talking about science geeks, of course! There are so many advantages to dating science geeks that it just surpasses my understanding why women aren't more enamored with us.

In the first place, they are generally available. Lacking in social skills, motorcycles and tight jeans they have been overlooked for so long that there is an over-abundance of them on today's market. Not only are they available, but another advantage in dating geeks is that other women seldom try to steal them. I can tell you from experience that my wife and I have been married for almost 50 years, and no woman has even made a pass at me. The only plausible explanation for this, of course, is that I am a science geek, and my wife has never allowed me to buy a motorcycle.

Geeks have other things going for them as well. One is that parents almost always love them. They appear harmless, often make good money and can fix things. That is no reason to marry someone, obviously. But it does remove some of the difficulties from life while you look over the passing parade of bass players.

Science geeks are surprisingly sensitive and romantic people, once you get past their initial, social awkwardness. For example, you wouldn't want to miss out on Valentine endearments such as:

The Rosette Nebula is red.
The Pleiades star cluster is blue.
The universe is expanding
Like my love for you!

It can be difficult to meet science geeks in the first place. They often have peculiar tastes in alternative music, so you seldom see them at concerts. They're even more rarely found in sports bars. They generally hang out in laboratories which have restrictive access, and they tend to socialize in groups where they discuss obscure and unintelligible topics. When seeking them out you can be at a distinct disadvantage.

But here's a tip. Guys wear t-shirts with logos of their favorite bands and sports team, thus showing that they are sinister dudes or are manly athletes, right? Well, science geeks tend to wear t-shirts with logos of software programs and science symbols emblazoned on them to show that they are, ah, well, geeks. Since there is a convivial rivalry about these things, you could try wearing one yourself. Try something like a nice tight T with slogans like:

-     I wear this shirt periodically
-     Never trust an atom - they make up everything
-     Code like a girl - or
-     Cole's Law = thinly sliced cabbage

    See if your tee strikes up any conversations!

    Of course the best way to meet science geeks is on the internet. Surfing the net allows science geeks to combine an activity with which they are comfortable - computing - with an activity they are uncomfortable with - socializing. Another strategy is to hang out in the junk food aisle of the grocery store.

    Most importantly, science geeks thrive on mystery. So just keep being female, and they will be helplessly fascinated - with an emphasis on the "helpless" part.

12. FUNDAMENTAL, INEXPLICABLE, ATTRACTIVE FORCES

Gravity got you down? (Sorry, Nerd joke.) But whether it does or not, there's not much you can do about it. Gravity is just there. It makes no sense to get mad about it. It's just one of those inexplicable, attractive forces.

Whenever humans don't know what something is, we give it a name, so we can talk about it. That's what happened when I was born. It worked so well that my wife and I did the same thing. We gave all our children different names too. Then we spent most of the rest of our lives talking about them. There are a lot of things that we give names to that we don't understand. . ..

Like, why don't I fly off a planet that is rotating at about 700-900 miles per hour? Logic says we should fly off, but we don't. So man invented a force that keeps us on and named the force gravity. We can sort of understand parts of gravity. We still don't know though, where it comes from.

By definition, gravity is the phenomenon by which physical bodies are attracted to each other. Gravity gives us weight and mass, which are not necessarily the same thing. Weight is the measurement of the pull of gravity on an object. We know that weight can change with location. Our weight on earth is not the same as our weight on the moon.

The mass of an object, however, is a measurement of the amount of matter something contains. If it contains more atoms it has greater mass, but its weight can still change with location. A heavy person can change their weight by going to the moon, but a dense person cannot change their density. You know what I mean. Anyway, gravity is one of the inexplicable attractive forces of nature.

Magnetism was discovered by Sherlock Ohm. (Are we paying attention?) There are numerous theories describing how it works based on opposite poles and attractions. Yet no one has a clue why there are opposite charges in the first place. Electromagnetism, which is what we named this force, is the force that causes the interaction between electrically charged particles. It explains most of the common phenomenon of everyday life. The particles that make up our world often have one of two opposite charges. Opposite poles attract each other and like charges repel.

These opposing elements are called charges in electricity, and poles in magnetism. An electrical current running in one direction creates a circular magnetic field around the wire perpendicular to the wire. In this way, magnets can be converted to electricity and electricity to magnetism. It's another one of those inexplicable attractive forces.

The attraction between men and women, while not generally recognized by the physics community as a "fundamental, inexplicable, attractive-force", meets all the above criteria. It has been much studied, there are innumerable theories, but the fundamental causes of this attraction are inexplicable. Our understanding is far less precise, and prediction is nearly impossible. Control? Fagetaboutit!

The physicists don't like to bring this whole topic up. They know that if the word gets out that the real serious scientific questions are about sex, they will lose disciples quickly. I mean, look around. Notice who good looking women are seen with and tell me it's explicable. (I think inexplicable in this context means "expletive deleted".)

We haven't even gotten to the real mystery. Everyone knows men are attracted to women. But women seem to be attracted to men. Try and figure that one out. I told my daughters that men are highly overrated and that I should know, being one. Alright, I admit they didn't listen, but you can't say I didn't warn them.

Here is the miracle, though, that shapes more men than our culture wants to recognize. It's the miracle that many young girls have never seen, and so, cannot comprehend. A man being attracted to women is one thing. That is biology and can be explained. A man being attracted to a particular woman is something else. That is something very different, perhaps almost holy. That one there, in the sweater, smiling shyly. . .. I must have that one. Happy Valentine's Day, Honey!

13. FALLING FOR SCIENCE

Love is the only thing that I can think of that a person can both "fall into" and "fall out of". It doesn't make any sense, but then you probably already knew that. We sometimes think of love like a thermometer - going up and down in varying degrees. Yet we fall either way. Love is peculiarly peculiar. Have you ever tried to say "peculiar" out loud several times in a row? It is peculiarly difficult to do.

Falling in love is peculiarly difficult sometimes. The first requirement is a person to fall in love with, and they are not so abundant. Personally, I believe it is impossible to fall in love with humanity. The best I have been able to achieve so far is "falling in tolerance" with them. I suppose loving humanity is probably a worthy goal. Politicians claim to be good at it.

As a young man, I had an interest in the feminine part of humanity. But I wouldn't have called it love. No, it was only when a specific woman came into my life that something, I call love, developed. I remember my buddies called it obsession, but they just didn't understand. Also, it wasn't so much like a fall as it was a delightful downhill stroll, along a dangerously narrow path, by a terrifying cliff face.

This valentine introspection caused me to wonder why I fell in love with science. It seems to me that scientists are peculiar in that way; we all seem to love our work. I did a lot of other things on the way to becoming a scientist, but I didn't love them. I love being a scientist.

With further thought, I don't think scientists fall in love with "science". Instead I think they fall in love with an object. It seems to me, from the scientists I know, that they fall in love with a radio, Legos, a plant, animals, computers, marbles, explosives, fire, gardens, models, or some other object that creates a fascination in them and allows them to manipulate their world.

A computer scientist I once heard speak, was a woman who, as a girl, became fascinated with braiding her dolls hair. From three strands to four, to six, and more strands of braiding, she claimed she was developing the idea of recursion long before she learned to program. Recursion is the use of a rule, of some kind, that calls for repeating itself several times before moving on to the next step.

Unfortunately, today's education doesn't have much time or space for people interested in objects. By the fourth or fifth grade, making things or tinkering, is eradicated from the curriculum. Those with such interests must usually find their own outlets and are often regarded as nerds, or something else other than normal.

The study of "science" didn't come into existence until about five hundred years ago. That was when people began to look for the general, abstract rules that controlled physical matter: objects. This is one kind of definition of science.

And the study of science only occurred after several thousand years prior, when humans were greatly occupied with turning abstract ideas into objects. Generally, people had an idea for an object they wanted to make, either for the purpose of increasing functional use, or often religious worship. Sometimes this approach to making objects is called "art" or "craftsmanship".

Bezalel, who fashioned the gold for the Ark of the Covenant, surely preceded the periodic chart of elements. The craftsman who learned to create precision water clocks fed the discovery of the speed of light. It seems like the arts were the chicken that laid the scientific egg.

Anyway, it's obviously my parents' fault that I love doing science. What did they think was going to happen with all that junk in the garage and seventeen different kinds of animals around? I fell in love with science, and I can't get out.

11. GUNS AND ROSES

I've always wondered if it was an accident that Horace Smith and Daniel Wesson patented their firearm on Valentine's Day of 1854. Were they trying to make a statement or something? Seems odd! I've even wondered if maybe they invented the Smith and Wesson before Valentine's Day was a thing. But apparently Valentines goes back a long way, at least to the third century.

Everything is odd about Valentine's Day. It is celebrated annually on February 14, and originated as a Western Christian feast day honoring an early Christian martyr named Saint Valentine. Can you believe a day devoted to love and affection would begin with someone dying? I think that is why it isn't a real holiday anywhere in the world. I mean, who wants to celebrate martyrs, whether it be for the love of God or the love of someone else?

The first recorded association of Valentine's Day with romantic love is believed to be in the *Parliament of Fowls,* a story by Geoffrey Chaucer written in 1382. You may have forgotten about Chaucer; or if you're one of the younger generations, you may never have heard of him. General education has changed considerably. Old people used to have to read his stories and sometimes even read it in old English.

Anyway, this story was supposed to honor the first anniversary of the engagement of fifteen-year-old King Richard II of England to fifteen-year-old Anne of Bohemia. Anne died at a relatively young age of the plague. So, Chaucer wrote the following, presented here in modern English.

"For this was on Saint Valentine's Day

40

When every bird comes there to choose his match
(Of every kind that men may think of!),
And that so huge a noise they began to make
That earth and air and tree and every lake
Was so full, that not easily was there space
For me to stand—so full was all the place."

They just don't write stuff like that anymore; probably because humans have killed off most of the birds. And they seldom get together in a convention like this. Very similar things still happen, though, at human conventions in the modern world.

Roses seemed to enter the picture as early as 1590 when Edmund Spenser wrote his epic poem *The Faerie Queene*.

"She bath'd with roses red, and violets blew,
And all the sweetest flowres, that in the forrest
grew."

It isn't entirely clear why roses, especially red ones are symbolic of love and affection. This association seems to go back even further than the martyrdom of St. Valentine. In western culture it has been believed that this particular flower was created by the goddess of love, Aphrodite. According to legend, her tears and her lover's blood watered the ground from where the red roses grew. Theirs was then a symbol of love until death. Yeah, you just knew it was going to end in tears or blood, if not both, didn't you?

The invention of gun powder just made the process more efficient. Smith and Wesson were merely another small step for man, but a giant leap for lovers. I mean, we quickly got from Tom Dooley and messy knives to Frankie and Johnny who were lovers! She shot him with a big forty-four. Probably a Smith and Wesson.

You may think I sound jaded. In fact, I have been greatly blessed. My wife and I have been married for over fifty years. Still, it is inescapable that - as we walk - you should stay behind. It's bound to end in tears. One of us will go first, and a rose will grow where our blood and tears intertwine. We have the advantage of believing we will be together again after death. If we didn't believe such, we would be hopeless, and that would be a fate worse than death.

If you love, know that you will either bleed or cry, and perhaps both. But, without the blood and tears, life would be lonely and worthless. So throw caution to the wind. Embrace the inevitable and celebrate Smith and Wesson Day every February 14th!

ABOUT THE AUTHOR

Dr. Gary L. McCallister is Professor Emeritus from Colorado Mesa University. That means he no longer professes anything at said university. He professed things biological at said university for over forty years. He is a highly trained professional windbag and can prove it with sixty-two scientific papers, eight books (so far), and an award-winning weekly science column for the newspaper. He has also produced 15 music CD's and is a luthier of the mountain dulcimer. His unique designs are popular throughout the western United States. He lives in western Colorado where he tends bees, grows prickly pear cactus, and plays music on his guitars, mandolins, banjos, and mountain dulcimers. You can contact him at: gmccallister@bresnan.net

OTHER BOOKS BY GARY MCCALLISTER

MUSIC
Making More than Music 2014
First Songs with the Mountain Dulcimer: history, instrument, and simple songs 2015
Hymns on Mountain Dulcimer: Learn to play the mountain dulcimer using hymns 2016

SCIENCE
Hanging Out With GRAVITY: Galileo's gravity game 2015
Seriously Silly Science: A science reader for the whole year - and some of it is even true 2015
A Convenient Truce: A cease fire in the war between religion and science 2016
The Solar Solution: the solution to problems you didn't even know you had. 2017
Between Two Mirrors: art and science in the modern world 2017
Science is Serous: all the scientists say so 2018
Thou Shalt Make: the spiritual significance of making things 2019
Reality: the cure for poor behavior 2019
A Simply Scientific Christmas 2020

NOVELS
Walking Man 2015

HISTORY
Grandpa Mac 2017
The Hammars of History 2018 (Editor and publisher)

All available on Amazon.com